♥**For Little Cat**

contents

A cat's perspective

Balancing act
A scamper over the picket fence, a quick dash up the tree, and with a sure-footed leap, the cat reaches dizzying heights. This is all just child's play for him; a true acrobat with an amazing sense of balance.

Best friends
A clear demonstration of friendship: a pair that understand one another. Sociable cats can form close friendships, greeting one another happily, rubbing against each other, and communicating with a variety of fragrant pheromones!

Foreign language friend
Even if dogs and cats don't speak the same language, they can still understand one another, especially if they have grown up together. Bonnie, the brown Labrador, and Tinker have formed a deep, mutual bond.

Pounce!
A young cat practices his hunting techniques and tactics.

Sharing the workload
While her owner sits at the computer, this cat takes the opportunity to relax. Cats spend a huge amount of time sleeping – up to 20 hours a day. They especially love stretching out in the sun or snoozing on the sofa.

Interesting scents
The great thing about feline scent-marking is that the scent hangs around long after the cat has gone. With urine, for example, the cat can stake out his territory, or discover where another cat is or what he has been up to, without having to meet.

It's all mine!
Cats rub their chins and cheeks on everything that has significance to them: boundaries, chairs, people. They secrete a scent from their scent glands, and are able to recognise their own when they encounter it again.

I'm the biggest!
The back is arched, the legs are stretched out, and the fur stands on end. This little kitty-cat wants to appear huge, even if he doesn't feel particularly confident.

Get out the way!
Bad mood alert! The whiskers are fanned out, the eyes narrowed into small slits, and one ear tilts backward. Get too close and you could be in for a swipe!

Engraving
Claw-sharpening leaves a clear message: 'I was here!' Not only do the claws get sharpened, but a clear visual signal and a scent are left behind. The paw pads have glands which secrete the scent as the cat scratches.

A cat's lick
Our domestic tigers spend a lot of time on hygiene and don't leave any part of their body unwashed. Their tongue acts as a comb. For the bits they can't reach with their tongue – the face or behind their ears – they lick their paws and rub them on these hard-to-reach areas.

A tense atmosphere
A feline punch-up: in this difference of opinion, the cat lying on his back has all four paws free, ready to swipe.

1

cat-speak what is it?

Enigmatic creatures: The secret lives of cats

Extraordinary individuals

Cats are not very easy to figure out. As hard as we may try to understand the meanings and motives behind their many different ways, a lot of what they do, and why they do it, remains a secret. Why is this?

Firstly, cats are all individuals. What one cat loves, another may hate; what one cat may tolerate with stoic serenity may make another cat furious, immediately indicated by a hiss and a swipe of the paw.

Precious independence

But that's not the whole story: these purring little moggies are far too clever to give us two-legged creatures a proper view into their inner world; nor do they want to divulge the reasons behind their mysterious feline behaviour. This would take away their aura of mystery, and that's the last thing they want, as their independence is sacred to them.

However, by translating your cat's 'language,' this book will help you reach a greater understanding of the everyday behaviour of your furry friend.

A little pleasure-seeker

Most cats love being massaged or

Cats are completely free spirits; they don't have a fixed routine, and come and go as they please.

One of the many reasons why we love them: what could be nicer than a cat who comes and goes of his own free will, cheekily invites himself into our home, and waits for us to make a fuss of him?

fed a tasty morsel – how about an extravagant liver and tuna creation? Cats can be very charming and can have you at their beck and call before you even realise what's happening. People have always got along well with cats – at least in a domestic set-up, anyway.

Communication with the 'two-legged creatures'

Cats seem to be fascinated by people, and are very keen to get along with us. In order to get to know our habits, interpret and understand us, cats are attentive and observant, despite the fact they have other priorities in their lives, and live in a completely different world when it comes to senses. But do they really need our hospitality, when all is said and done?

What does 'miaow' mean?

Why have cats evolved a rich vocabulary of sounds in order to communicate with us? 'Miaow' has many meanings, and is usually directed specifically at people. For verbal communication with other creatures, cats use a completely different set of sounds.

A great affection for the feline species

Understanding one another's language is only one requirement for a harmonious relationship between cat and human. A deeper understanding of the domestic tiger and his specific characteristics, as well as his behaviour, is also necessary.

Cats are fiercely independent predators, and only do things on their terms. It's never nice to find the body of a bird or mouse that your cat has killed, but it is vital to understand that cats are hardwired to hunt; millions of years of evolution cannot be denied.

Paws for thought:

When cats sleep

Although cats are very active at dawn, dusk and during the night, and then sleep for more or less the whole day, some are able to adapt their circadian rhythm to that of their owner.

Cat naps and the champion snoozer

Cats are world champions at sleeping, and sleep or doze for around 16 hours a day. If it's warm and comfortable, and your furry friend is well-fed and feels safe in his surroundings, he will sleep even longer – up to around 20 hours per day.

Possibly a record breaker ...

New-born kittens and very old cats sleep an extra two hours on top of this, so the time they spend awake is considerably shorter. In the case of kittens, this changes rapidly; by the age of four weeks, they already sleep for the same amount of time as an adult cat.

Real dreamers

Cats not only sleep for a great length of time, they also dream quite a lot, too. It has been discovered that their dreams last for up to three hours a day. Every 25 minutes, the so-called non-REM phase (light sleep) is broken by a phase where the animal is in such a deep sleep he is barely able to move.

In spite of the lack of muscle tension, his paws, whiskers, and even tail may start to twitch. Sometimes, even the nose twitches as well. His eyes roll back in his head and can be seen between the eyelid and the nictitating membrane (or 'third eyelid'). These rapid eye movements are caused by the dream phase known as REM sleep, and this occurs with cats and people alike.

Moving moment

Although a cat is fully immobile while he is in this dream phase, his brain goes into overdrive, sifting through and sorting out recent events. Interestingly, most dream phases seem to happen after a cat has been hunting, which is a very exciting and turbulent time for moggies. At any rate, the involuntary twitching of the face and paws, as well as the REM, show us that our cat is dreaming.

What our domestic tigers are

actually dreaming *about*, and what sensations this generates for them will remain their secret, of course ...

Cosy cats

A cat can nap almost anywhere, but, if intending to have a proper sleep, will seek out a warm, safe place where he can truly relax, and won't have to regulate his body temperature (during REM sleep, a cat's body becomes noticeably cooler).

Curled up or stretched out?

The position in which a cat chooses to sleep will largely depend on ambient temperature. If it's lower than 10°C, he will curl up and tuck his head into his body. As soon as he begins to warm up, he will stretch out a little, and if the temperature reaches over 20°C, he will stretch out completely.

Occasionally, cats will even sleep on their backs with their legs stretched up in the air!

Yawning

When a cat wakes up, he will stretch his body as much as possible in order to elongate his muscles and get the circulation going again. Next, he will stretch out his front legs and hold the stretch, also taking the opportunity to extend his claws and arch his back, stick his bottom in the air, and then stretch his back legs and tail.

Often he will yawn, and then begin to clean himself. Cats also yawn if they are unsure about something; in some cases, the yawn serves as a stress release.

A luxurious yawn doubles up as a muscle stretch after a nap.

For the kids:

Washing – a 'cat's lick'

Have you ever wondered why people say "That wasn't a proper wash! That was just a 'cat's lick'" when, in fact, a cat's lick is actually far more effective than most people realise?

Extensive daily wash

Cats wash themselves every time they wake up. A quick lick and a rub of the paws over the face is just the start; they are extremely thorough when it comes to daily hygiene. In fact, researchers have discovered that cats wash themselves several times a day, spending a total of two hours doing so. This is a considerable amount of time, and can hardly be regarded as "just a cat's lick"'!

"That was a bit embarrassing ..."

Cats scratch, lick, and nibble, not only as part of their cleaning ritual but also, for example, when they are unsure or nervous, just as we would scratch our heads or fiddle with our hair. These actions are known as 'displacement activities,' and are only a superficial attempt at cleaning, which could explain why some people think that cats are neglectful of their personal hygiene.

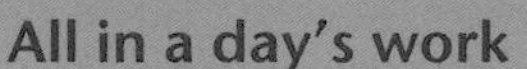

All in a day's work

Watch how expertly your cat cleans himself: a task he carries out with great care, leaving no part of himself unwashed. He uses his tongue to do this, and stretches his flexible body this way and that to ensure a thorough job is done.

No grubby urchins here!

An early start. At the age of around three weeks kittens begin to learn how to clean themselves; by the time they are six weeks old, they have perfected this ritual.

Be careful: I'm very preoccupied!

When a cat cleans himself, it can seem as though he is in a trance. Should you try and stroke or touch him during his wash, your hand may be either licked or unceremoniously bitten, depending on your cat's temperament, so be very careful!

A world of their own:

How cats perceive their surroundings

Be they hearing, seeing, feeling, smelling, tasting, or using their amazing sense of balance, cats are always one step ahead of us

Little eavesdroppers

Not only can they hear extremely quiet and much higher frequency sounds than us (their hearing capabilities reach well into the ultrasonic range), cats are also better at differentiating one sound from another.

The reason for this is that cats' ears are specially designed to hear minute differences in tone, and the smallest variation in volume over huge distances. Because cats use their impressive ear muscles to locate noise, they can hear sounds from a much further distance than we can. Also, our furry friends can accurately identify different noises, and pinpoint a particular sound from a huge range of high level noise.

Not only can cats hear frequencies three times higher than we can, they can also hear sounds that are three times quieter.

No wonder, then, they can 'read' us so well because, by listening to the different tones of our voices, they can tell what mood we're in.

Cats' eyes

Not only can cats hear better than us, they win hands down when it comes to eyesight, too. Their eyes have a considerably higher sensitivity to light and movement than ours. Admittedly, cats don't see as sharply as we do; they are not as good at recognising colours, and they can't perceive optical depth as well as we can, so their spatial vision is not as good as ours, but this is no loss to them.

We presume that cats see their surroundings as one colour, and use other senses for orientation or catching prey. As well as their incredible acoustic and visual perception, they have sensitive whiskers, and an amazing sense of balance.

I can read you like a book

Not only do cats have an excellent perception of their surroundings, they can use this sense to communicate within their species. When a cat, for example, locates a certain sound, his ear muscles will move in rapid succession. Based on this reaction, it is possible for other cats to notice what their fellow cat has seen, how strongly his interest has been piqued, and whether it is worth taking a look themselves. If we carefully observe our cats, we too can learn to 'read' their motives and intentions, just by looking at the position and movement of the ears, the direction they are pointing in, the size of the pupils, and the movement of the head.

A shared perspective

Certainly, we two-legged creatures are not as good at reading our cats as they are at reading us. Often, each individual action is too varied, and one too rapidly followed by another. The human eye is simply overwhelmed by this, which makes interpretation very difficult.

With a digital camera take a quick sequence of images and you will be able to capture what the naked eye cannot easily do. Also, a small dictaphone can be useful to record the tone and volume of your cat's voice. All of this information can then be put together to enable you to build a clearer picture of what your cat is trying to say.

In order to respond appropriately, we must first familiarise ourselves with feline language and vocabulary, and we take a look at this in the next section.

Behind their retina, cats have a residual low-light amplifier called the tapetum lucidum, which allows them to see and hunt successfully at dawn and dusk. This is also the reason why cats' eyes seem to 'glow' in the dark

Up close and personal.These cats are familiar with one another, and make contact by sniffing and rubbing against each other.

Little chatterboxes: Feline vocabulary

The plaintive, high frequency 'miaow' of our beloved moggy makes us come running!

Communication between cats can be in the form of sounds, facial expressions, touch, body language, movement of particular parts of the body, or through a sequence of different behaviours, most of which are accompanied by scent-marking.

When cats communicate with each other, body language and scent have a more significant meaning than sounds. It is only with us that cats 'chat' a lot, using a multitude of different sounds and noises

Cats also have an extensive repertoire of 'conversation' at their disposal, which they use to express their moods, wants and needs. They do use this language with other animals, but are more likely to use it with people.

An experienced translator

Which 'vocabulary' a cat uses in each individual case, and how successful the transmission of information actually is, depends not only on the experience of the receiver, but also that of the cat.

Cats who are used to being surrounded by people communicate considerably better using loud utterings than those who have had far less contact. Through the experience of living with people, they learn how important the spoken word is to us. This ability to communicate doesn't depend on the breed of cat, even though each breed differs significantly in terms of conversational style.

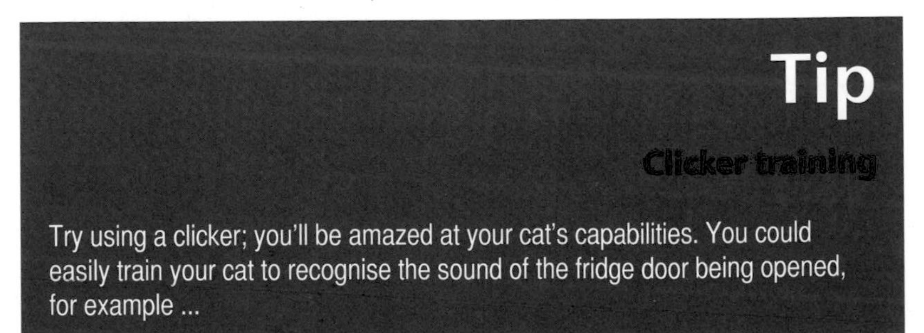

Tip

Clicker training

Try using a clicker; you'll be amazed at your cat's capabilities. You could easily train your cat to recognise the sound of the fridge door being opened, for example ...

Chatting with dogs

When dogs and cats come across one another, a cat who is used to dogs will communicate using facial expressions and body language, just as he would do with one of his own species. Cats clearly use less 'vocabulary' to communicate with other species than they do with 'speech-orientated' people.

Lost in translation

It is common for cats and dogs to misunderstand each other. A physical signal – for example, a vigorously wagging tail or a raised paw – can, unfortunately, signify the opposite meaning to each animal, and in the best case scenario would elicit a surprised response. In time, dogs and cats can learn to tolerate one another, or even get along well, in spite of their differences.

Learning a foreign language

Cats are fascinated by what people do and say, and more importantly how we say it. Even if they cannot understand the exact words, they learn to associate recurrent noises (the same applies to the quality and frequency of the tone) with a particular behaviour or object.

Behavioural experts have discovered that cats are able to differentiate between and understand around 30-50 words; no mean feat for a supposed loner who actually requires no intense social interaction. The fact that a cat tries to understand our strange behaviour as well as our language shows that he probably has a stronger social element to him than we realise.

This young dog knows the appropriate way to behave around cats, and receives a typical moggy greeting in return.

'Miiiiaow!'

The words which cats pay attention to the most are those with an 'e' sound in them, especially when they are spoken in a high-pitched voice. They also show interest in high-pitched squealing noises, but that's no surprise when you consider that their favourite prey, the mouse, squeaks at a high frequency.

Out in the open: better to stay out of the way or zero in on one another? The mutual stare is not a good sign.

2

do you understand me?

More than a thousand words: 'Miaow' has many meanings

Cats listen very carefully to human speech, and, by paying attention to our intonation when we talk to them, sometimes reward us with more than a simple 'miaow' in return

An innate vocabulary

Unlike people, who only learn to speak by mimicking what they hear, cats don't learn by example.

Cats who are born deaf still learn their 'spoken' language in all its varieties without ever actually hearing any sounds that they can copy. Cats are born with an innate ability to miaow, purr, and hiss, and although their repertoire is not perfected straight away, through trial and error, they soon discover what works, and which sounds are suitable for each and every situation and occasion.

The all-encompassing miaow

Similarly, our cats try to figure out our reaction and then decide specifically which type of 'miaow' would be the most effective. They will test out the location, the context, and different sounds on different people. A 'miaow' is usually a request for something that a cat requires from us. Adult animals seldom use this miaowing sound to each other.

It's all in the intonation

The miaow can be altered very effectively. Every syllable can be stretched out using varying intonation,

Even the tiniest kitten soon begins to try out his little voice.

"Hello little kitty" 'Miaow' (he looks straight at you).

"How are you today?" 'Mew mew mew' (he waves his tail in the air).

"Come here, then" 'Miiiiaow' (he purposefully makes his way to you).

"Tell me, how was your day?" 'Miaow-purr-purr' (he pushes his head against you and starts to purr).

and with each slight alteration, the meaning is changed. If a cat is disappointed, more emphasis is put on the 'ow' sound in 'miaow.' Sometimes, only the 'ow' is uttered rather than the whole 'miaow.' If the 'ow' has a higher intonation, the noise sounds a bit like distress. This means a cat is pleading for something. If the pleading pays off, the 'ow' is uttered again, and the mouth closes very slowly so that the message lasts for longer. If a moggy is in a good mood, the 'miaow' is not as intense and sounds lighter, sometimes followed by a purr.

Miaowing in ultrasonic range?

The ultrasonic range of a cat has never been fully explored. It is thought that the so-called 'mute miaow' is not actually silent. Instead, it is at such a high frequency that because of our limited aural abilities, we simply can't hear it. You're bound to recognise this gentle pleading miaow: your cat opens his mouth wide and then shuts it again, without actually making any noise. But if you were to put your head against his mouth while he does this 'silent' miaow, you would hear a little whisper. Other cats with their acute hearing would definitely not miss this form of communication. It could be that these animals actually communicate verbally with one another more than we think they do, but in a different ultrasonic range: this would be a useful communication tool for cats. So the little moggy with a social streak may stay in closer contact with his feline friends than previously assumed.

Cat-chat: When moggies miaow, coo, twitter and sing

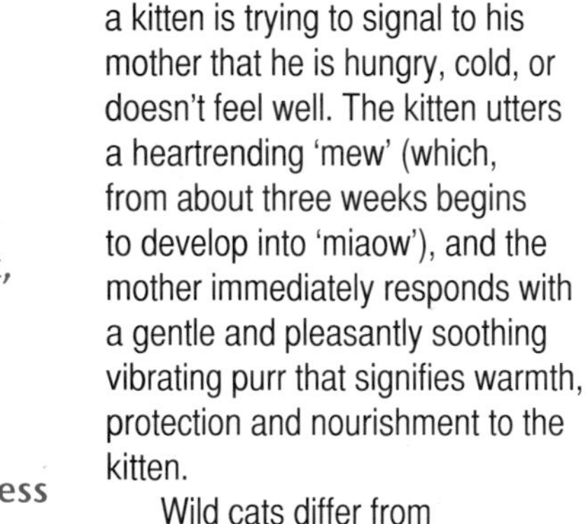

Tentative or urgent, softly cooing or mournfully echoing: cat-speak can express a full range of emotions

Whenever cats want something, they miaow; for example, when a kitten is trying to signal to his mother that he is hungry, cold, or doesn't feel well. The kitten utters a heartrending 'mew' (which, from about three weeks begins to develop into 'miaow'), and the mother immediately responds with a gentle and pleasantly soothing vibrating purr that signifies warmth, protection and nourishment to the kitten.

Wild cats differ from domesticated felines because they lose this communication as soon as they reach maturity, whilst our house cats retain this early kitten-like behaviour long after they have fully matured. This is because they are in contact with people who respond to their 'miaows,' so they learn to modify and adapt these sounds for each different need.

Cat dialect

Whether they are pleading or complaining about something, every cat has his own 'miaow dialect.' Some animals chatter away very loudly and enthusiastically, whilst others introduce cooing and twittering into their cat-conversation with gentle mews, in order to have a friendly chat with their humans. It is not unusual to hear the affectionate 'rau rau rau' sound from a happy and contented cat.

Gentle tones

The mother (queen) emits gentle cooing and twittering sounds when she returns to her young, and a somewhat louder

On hearing a call from his mother, the kitten comes running and is greeted lovingly.

'trilling' tirade when she is trying to encourage her kittens to do something. If the kitten is slightly older and a little more independent, the mother calls to him in this way to encourage him to feed. Tom cats often court female cats with this 'trilling' sound, used as a kind of chivalrous small-talk. If the female is interested, she will answer in the same way, with these sweet nothings possibly lasting for an hour or more.

Amongst tenors

So-called 'caterwauling' makes for a very impressive show. It's actually not a joyous song, but the result of territorial and ranking disputes between tom cats. In spite of their solemnity, the loudly repetitive 'mau mau mau mau' performance sounds more like a small child who needs his mother than the typical macho behaviour used in a mating ritual. But the caterwauling is deadly serious, as shown by the menacing body language which accompanies it. Incidentally, the 'song' is not always performed in the presence of an adoring female.

Making a racket

Occasionally, without a female nearby, the tom cat 'sings' in this

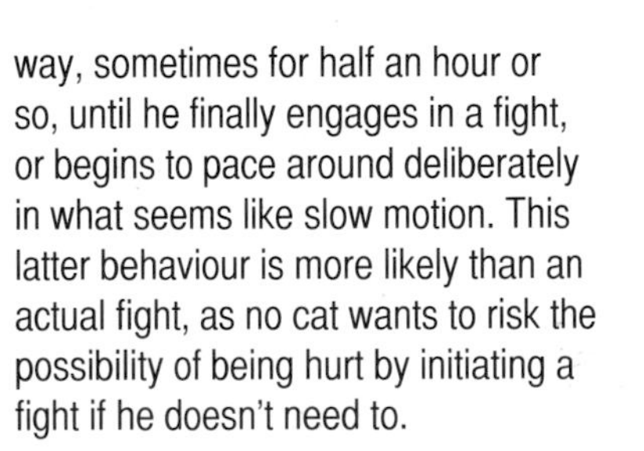

way, sometimes for half an hour or so, until he finally engages in a fight, or begins to pace around deliberately in what seems like slow motion. This latter behaviour is more likely than an actual fight, as no cat wants to risk the possibility of being hurt by initiating a fight if he doesn't need to.

The experienced courting cat, Jerry, knows the score: waiting patiently before beginning his romantic singing soon gets results.

This is serious: Spitting, hissing and growling

The beginning of a bust-up: the young cat refuses to accept his mother's telling off, and decides to rebel.

Cats can emit a bloodcurdling wail to make their fury known. But this wail can also be articulated in many different tones, some much softer than others

Sometimes, an unsuspecting person can be confronted by an unpleasant wailing sound from a cat, instead of a coo, twitter or 'miaow.' For example, it could be a hissing or spitting sound, if he or she has missed or ignored a previous warning signal.

Hissing

Hissing goes something like this: the cat's face will tense and wrinkle up, with the brows furrowed, the mouth open halfway, the upper lip pulled back, the teeth bared. The tongue is placed up near the top row of incisors, and a short, sharp blast of air is quickly exhaled. Anyone standing near enough would be able to feel the puff of air coming from the cat's mouth. The action of hissing alone can warn off an unwanted intruder. When the air is harshly and slowly exhaled, the sound then mimics the hiss of a snake, which is even more daunting. The hiss is used as a defence, a forewarning of attack, but can also be a sign of uncertainty, fear, or rage. It is usually aimed at another cat, but can sometimes be directed towards a person, or even a dog.

Spitting

If hissing doesn't have the desired effect, a cat has another way to express his displeasure, and give his opponent a strong warning signal of imminent attack: spitting. The cat exhales a sudden burst of air, and along with this, a shocking, wailing noise which

'Dinner's ready!'

A softer, almost lamenting wail which sounds like a modified 'miaow' is made by the mother when she brings food home for her kittens, and wants to attract their attention. From this wailing sound, the kittens can identify which type of prey she has brought home. The so-called 'mouse-call' (a phrase coined by cat experts) makes the kittens come running, whereas the 'rat-call' elicits a less enthusiastic response.

This amorous approach is less than welcome, as evidenced by the female's expressions and undoubted vocal objection.

is designed to frighten the enemy and give him the opportunity to flee. In order to underline the seriousness of the situation, the cat will threaten a fight by holding either one or both of his forepaws out in front of him.

Grumbling and growling

If the spitting doesn't work, the cat resorts to the last weapon in his non-violent protest: growling. This is done with the mouth closed but the corner slightly curled up.

Long, drawn-out growling is an unmistakable warning sign of attack and potentially serious harm. Once the paws are up in the air, a fight is inevitable. The moggy is now hopping mad, and is hell-bent on ensuring his opponent backs off (for example, if he is defending a tasty morsel). The volume of growling increases, and this frighteningly effective sound means: 'Leave me in peace or I'll beat you up!'

Menacing wail

If a cat feels like he has been backed into a corner, he may emit a guttural, menacing wail, and may shriek as well. The shrieking sound is also heard when a cat is in pain (for example, if you unwittingly step on his paw!), and this noise is also used by the female once mating is finished, when the tom cat withdraws. Withdrawal causes her severe pain because the tom cat's penis is covered in tiny barbs.

Also, if a cat has located an intruder in his territory, an angry warning cry is let out, much sharper than the usual wailing noise.

Kitten training

If a mother hisses at her kitten, this means: 'Watch out! That's dangerous!' or 'Will you behave?' Older siblings or adult cats will sometimes grumble or growl and then start to hiss as a warning sign that the kitten's behaviour is unacceptable.

Purrfectly content: Purring

Cats can purr for hours at a time. Sometimes the purring is soft and quiet, and sometimes it's loud and demanding; this is often the case if a cat feels frightened or ill

All's well

Purring is an unmistakeable noise that has a blissful, soothing effect. Even a tiny, week-old kitten is capable of purring, usually when he is suckling; when his mother hears this sound, she knows that everything is fine.

When a mother returns to her kittens, she purrs to show her little ones that everything is in order. Adult cats also emit an intense purring sound to show that they have nothing but friendly intentions. Kittens purr to older cats around them, and the older cats will purr in response if they are happy to be in close contact with the kittens.

Cats sometimes purr when they are eating or playing, which signals that they are in a good mood. Purring is (at least in this context) a customary means between cats to show that they are good-tempered, although only ever between cats in the same family or group. Puppies or dogs within the family often receive vigourous purring from the cat.

Purring in pain

Conversely, cats also purr if they are ill or injured, in pain, or even dying. Possibly, they are trying to demonstrate their weakness and appease a potential

A mother's caress and quietly vibrating purr slowly ease the tiny kitten's anxiety.

enemy by reassuring them that they are not a threat; they may also be trying to comfort themselves as well.

Completely relaxed

The cat purr is very appealing; what cat owner is not familiar with the cosy feeling that the gentle purring of their cat provokes? The peaceful thrumming of his purr, which slowly builds until his whole body resonates with it, has a very soothing and positive effect on our wellbeing, causing blood pressure to drop and instilling a lovely sense of relaxation. A cat's gentle snoring is also quite soothing, but doesn't have the same comforting effect that the purr does.

After a cat has been stroked, he will usually clean himself in an effort to mingle his feline scent with the human aroma left behind.

The mystery of a cat's purr

Cats are able to purr when they breathe in and out, and also while they drink, eat, suckle, or doze. Exactly how they do this is still not really clear. There are several theories, but none of them is wholly conclusive:

- It could be due to movement or 'turbulence' in the circulatory system

- The vestibular fold next to the vocal chords may begin to move, which causes the vibrating sound as the cat breathes in and out

- It could be the result of alternating contractions in the larynx and diaphragm

Fleur and Tinker enjoy snuggling up together in the sun, purring in unison.

I trust you:

Kneading, head-rubbing, and more besides

Without words: intimate contact between mother, Lilly, and daughter, Tina, is a sign of a close bond.

Kneading

Kneading is an expression of pleasure or comfort which is a throwback from kittenhood. As soon as a tiny kitten fastens his mouth on his mother's teat and begins to suckle, he simultaneously pushes his front paws against her stomach with his toes spread wide apart, and this increases the flow of milk through the teat.

Adult cats also do this with people (for example, when they're being stroked) as a way of displaying great affection, contentment, and wellbeing. During this rhythmic kneading, the cat's eyes are almost completely closed, and he sometimes appears to be in a trance. At any rate, he is quite obviously in seventh heaven!

Head-rubbing

The head rub is another sign of trust and bonding. You, as a cat-lover, are bound to know what is meant by this: your cat rushes up to you and rubs his head against your outstretched hand or your face with great relish, eyes shut all the while. During this ritual, he rubs his forehead, chin, cheeks and lips – and therefore his scent – onto your skin.

Stroking the top of his head, under the chin, across the cheeks, and over his back down to the base of the tail will further enhance the bonding process.

Rubbing against the legs

If your cat sees you with a tasty morsel or wishes to greet you, he will rub his body along your legs from his head to his tail. He may also wind himself around your ankles or calves, and raise his body in the air so that the tip of his tail connects as high up as possible. A cat who has absolute trust in his owner will sometimes lick or nibble the finger, back of the hand, or forearm of his human.

A further show of approval

A moggy's biggest token of love is the gentle 'eskimo kiss,' where his nose touches yours whilst you are face to face with him.

Transferral of scent

When cats rub themselves against objects or people, it's not just about the desire to make contact. Next to the taste receptors, cats have glands which secrete their own personal scent, and they use this to mark their territory, every (new) object within it, and also living things. Cats scent-mark to stamp their claim on their surroundings, and enhance their sense of belonging; the familiar cocktail of scent makes them feel secure and content whenever they smell it.

My scent, your scent

When a cat rubs himself against something, not only does he give off scent, he also acquires the scent of a familiar person or another cat who lives with him, as well as all the other smells of their surroundings. This means he will carry different scents on his fur, which he can pass on by rubbing against something else, and also inhale the aromas whilst grooming himself. This, again, helps him to feel secure in his surroundings.

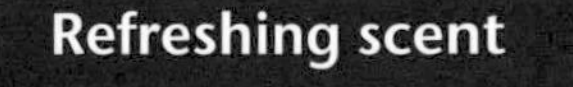

Refreshing scent

Although scent-marks are much longer-lasting messages than sounds or visual signals, they do disappear with time, which means that there's an instinctive need to scent-mark immediate surroundings as often as possible. Cats are thoroughly hooked on rubbing themselves against everything in their territory: a door, a chair, a hedge, a trusted member of their own species, and, of course, the people who 'belong' to them!

Elaborate rubbing and scratching on the fence leaves the message that 'Minky was here!'

Paws for thought:

Pheromones: mind-blowing aromas

Not just a grimace

If the scent signal doesn't quite reach the nose like other scents do, what then?

Just like dogs and horses, cats have a small, additional odour sensory area above the incisors called the vomeronasal organ, or the Jacobson's organ, where scent molecules latch on to sensory cells, causing a sort of 'scent impact.' And what an impact it is, usually sending a cat into raptures, although, with some, the effect is not quite as remarkable. It mostly depends on the individual.

Through the mouth and nose

Interestingly, the vomeronasal organ has two openings: one in the nostrils and one in the mouth on the gums.

This means that maximum stimulation of the senses is only guaranteed when the scent enters both the cat's nostrils and mouth, where the heavily-scented particles are released into the saliva and delivered to the receptor cells.

A curled upper lip

When a cat successfully supplies this 'smell-taste' organ with inebriating pheromones, this is called the Flehmen Response. The cat raises his head in the air, opens his mouth slightly, curls the upper lip, wrinkles his nose, forcefully inhales the air, and, with a quivering tongue, firmly presses the

A heather haven of scent: cats like to regularly deposit their scent, and investigate other scents in their surroundings.

Only for felines

Pheromones are like messages that are 'understood' exclusively between members of the same species. Cat pheromones can only be sensed by other cats; we can't smell them. We are also completely unaware of the strong scent 'flare' we carry around after we have cuddled our cat, but this could be why other cats are drawn to sniff us after we have been in contact with our own cat.

mix of air and pheromones against his gums at the opening of the organ.

It is only by making this distinctive 'grimacing' expression that a cat can acquire vital information from specific scent particles, which can tell him about the social or reproductive status of another cat, for example.

The Flehmen Response (which is when an animal curls the upper lip to facilitate the transfer of pheromones and other scents into the vomeronasal organ), is exhibited by all felines and particularly tom cats

Performing the Flehmen Response takes practice: at just four weeks old, this kitten makes his first attempt.

Sniffing, licking, tasting:

Smells everywhere

Oblivious contentment.

Pheromones are especially important during the mating season. Scent particles which are covered in pheromones have a very strong effect on feline behaviour. But it's not only adult cats that can smell pheromones.

A soothing scent

The newborn kitten is already familiar with his mother's 'soothing pheromones,' which create a relaxed and pleasant feeling whenever the kitten suckles her. This so-called 'Cat-Appeasing Pheromone' (CAP) is secreted from the mother's stomach after she has given birth, directly between the milk teats, where it is appreciatively received by her kittens whenever they are being fed or cuddled.

Young or old – the effect is the same

Although the mother cat only produces this soothing pheromone when her kittens are small and not yet weaned, it has the same effect on cats of all ages. When working with anxious or fearful animals, vets and behavioural therapists sometimes use an artificially created version of the pheromone.

A greeting in 'cat-ish'

It is very important for a cat to collect as many individual scents from other cats as possible. These interesting pheromone scents are not transported on the wind like other smells, and are only effective for a small distance. Cats must aproach one another and make contact in order to pick up on and analyse the pheromones.

As close as possible

Watching two cats meeting and greeting one another is truly a pleasure to observe.

The two press their noses against each other, and then sniff the nose and mouth, and sometimes the cheeks, and then up along the ears. Then out comes the tongue, and they begin to lick each other's face. After this short greeting, they rub the sides of their bodies and their tails together.

These young cats are the best of friends and two of a kind.

How long this actually goes on for all depends on how long ago they last saw one another. The longer the meeting lasts, the more intimate the body contact, and also the exchange of scents. Every now and again they can be seen sniffing the base of the tail, and occasionally the anal area.

Firm friends

After the greeting, the animals may walk together side-by-side for a short while to explore their surroundings, or cuddle up and share a morsel of food.

Intense body contact seems to be very important to cats, and those who are firm friends can cuddle up together for hours.

Cats recognise one another by smell; it is thought that the secretion of pheromones from the glands in their cheeks and chin deliver this information.

After a prolonged and intimate greeting, these two wander off together and go exploring ...

Scents rule the feline world:

Spraying and other scent-marking

As well as head- and body-rubs, cats have other effective methods of leaving their scent behind, and different ways of saying, 'I was here.' Spraying is very effective, as is territory marking with faeces.

Owners of 'intact' cats (those not neutered) may find that spraying is the more common behaviour.

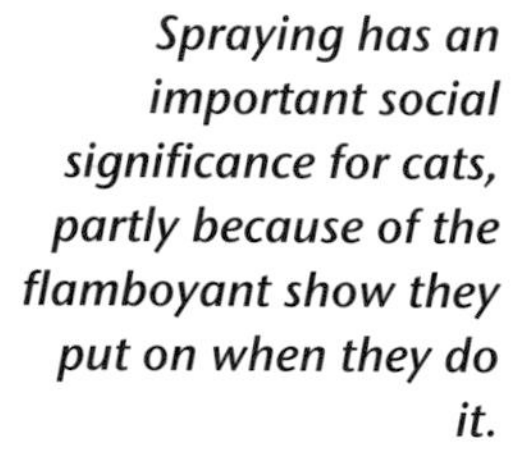

Spraying has an important social significance for cats, partly because of the flamboyant show they put on when they do it.

Spraying

The animal decides on a carefully chosen and conspicuous area which is usually vertical (for example, a tree), with a height of at least 30cm.

After thoroughly checking this fpr other scents, he turns around, places his bottom on the area and stands up straight with a conspicuously quivering tail. Spraying a small stream of urine onto the chosen patch, he may then use his hind legs to cover what he's done.

Even to urinate, a cat will painstakingly seek out the perfect place, and sometimes even cover over his urine.

The human nose can detect it, too!

Some tom cats also add a few anal secretions to intensify the scent, and if this process is repeated often enough, it can also be detected by the human nose. This scent-marking – a mixture of concentrated cat urine as well as anal gland secretions – is full of pheromones which stay burrowed in our nostrils. These pheromones contain millions of scent particles which are registered in the olfactory glands in the nasal cavity of cats as well as people.

A scent banner

Much like colourful flags would do, the sprayed scent-marks waft in the air in and around a cat's territory, and are particularly obvious on boundary lines.

It's not only tom cats who spray, either, as neutered and spayed cats also do this when they want to express their social status. However, it tends to be the dominant animals who lay a claim on their territory on a regular basis.

My territory!

The more cats there are within a defined area, the more regularly they will spray. Unfortunately, house cats will also do this, for example, if strange animals move in, or if there are any obvious changes to their surroundings. If you move house, your moggy may attempt to scent-mark anything and everything he can. Even a new wardrobe or other similar furnishings will, in a cat's opinion, need to receive a personal scent-mark in order to be permanently accepted into his living quarters!

Anal scent-marking

Cats (and this usually applies to cats who haven't been neutered) will sometimes mark their routes and territory with faeces. In this case, their deposits are not painstakingly buried, but instead left out in the open, usually mixed in with little clumps of soil or other light materials such as sawdust. This means they can be seen – and smelt – by all.

Burying faeces and/or urine masks the smell. Evidently, only a few animals with a high social status spread their scent in this way; less dominant animals prefer to use more subtle messages such as pheromones and scratch-marks (see page 42). Subordinate cats also bury their faeces.

The most elaborate scent-markings are deposited when a cat defecates. He finds a suitable place and digs a hole in which to defecate, after which he first sniffs and then covers the faeces.

No sweat!

Panting helps your cat to stay cool

Cats seldom pant, and if they do, it's not as effective as a dog's pant

It's not only dogs and birds that pant: cats do, too, if they get too hot. But how?

When cats pant, they breathe quickly in and out, and give off moisture; this causes them to lose a large amount of saliva and also heat, which helps cool them.

A necessity of effective panting is that the cat is well hydrated. Cats are unable to sweat in the usual way, and the feet are the only areas that contain sweat glands.

A cool place

As well as panting – which only occurs when it is unusually hot – cats have developed several other strategies to keep cool. They avoid lying in the sun during the hottest hours of the day, and instead choose a cool place to doze; they roll around and stretch out in damp foliage or soil; they lick their bodies extensively and fluff out the hair on their back, stomach, and flanks, which causes evaporative cooling, so that they don't need to pant.

Be aware, however, that regular panting can be an indication of a serious illness.

The cat's nose – a cooler for the brain

Particularly fascinating is the miraculous

Lilly rolls around in the damp earth to cool herself.

network of blood vessels in a cat's brain that helps keep him cool, a multi-branched, vascular arrangement which ensures that the delicate cells in the brain do not overheat and get damaged. The blood circulates this clever cooling structure, and when it has been cooled, it is carried by the blood vessels to the nostrils. Then, with each exhalation, any excess heat passes out of the nostrils and into the atmosphere.

Not every species of animal has this marvellous ability to cool their brain, but the few that do (cats and dogs, for example) are able to keep a cooler head in extreme heat.

Ways that cats deal with heat: Licking and grooming the limbs (far left) ...

... fluffing out the hair along the back, stomach and flanks ...

... and by panting.

Claw-sharpening or something else?

Language of the paws

A manicure for cats

Scratching is an undesirable behaviour when done inside the house on furniture and furnishings, yet it is entirely natural conduct for a cat.

Cats who claw trees or other wooden objects are not only sharpening their most deadly weapon, they are also removing the brittle outer layer of the claw, (sometimes these light layers are completely stripped off), honing the claws to the right length, and cleaning dirt from the underside.

Horizontal or vertical surfaces, hard or soft wood, cats will use whatever suits their needs.

Leaving their mark

Almost as important as claw-sharpening is the sight and smell of the obvious traces which cats leave behind after their 'manicure.'

These signs are visible to other cats, and are received as messages in a similar way to scent-marking. The intensity and length of the scratch leaves a clear visual signal about which cat has been there. The more self-assured a cat is, the more obvious the scratch marks are, and these dominant cats will regularly scratch in prominent areas.

Look over there!

Cats sometimes put on a bit of a show when they scratch. It has been noted that they carry out this behaviour more flamboyantly when in the presence of other cats, because they know they are being watched. The clear message

of this performance is, 'Watch out, my friends, I'm the boss around here!!'

More 'scent information'

A scratching session is also used as a scent-marking exercise. The scent is transmitted through the sweat glands in the paw pads, and is released when a cat drags his claws against an object. The only place a cat has sweat glands is in the feet.

The scent is not only deposited over the scratched area, but also secreted into the grooves or claw-marks, where it stays for a long time, the message left by the scratching evaporating slowly into the air. The deeper the grooves, the more space there is to spread the scent, and the more powerful the message.

Sweaty feet

How well the sweat glands work – meaning how much of a scent is given off – depends on the state of physical arousal in a cat. The stronger the arousal, the stronger the secretion. If it is especially hot, or if a cat has a fever, he will secrete more scent, in which case, it's more about cooling himself than scent-marking.

In order to keep his paw pads sleek and supple (which, amongst other things, is necessary for touch and successful hunting), a cat always has hot, sweaty feet! So every step leaves scent-marks behind in the area, which act as a kind of 'scent flare' to other cats

Even we two-legged creatures with our amateurish sense of smell are able to detect this musky sweet secretion on the soles of our cat's feet, but only if we were to put our nose close to them. Whether we smell the scent in the same way that another cat would is questionable. But it's certainly an interesting smell, all the same!

The cat receives a manicure when he scratches his claws on a tree. Cats grip onto the scratching area, spread out their toes, and drag their claws through the wood with great relish.

Flick knives: Cat claws in action

What makes a cat's paws so fascinating? One minute they are soft and velvety, and the next as sharp as a knife. The reason for this, as we all know, is that cats can unsheath their claws at lightening speed on demand, then tuck them away again as if they were never there at all.

The claws are a deadly weapon ...

Show me your claws

Because the tendons attached to the toe joints are so springy, the claws are easily extended, so that they can grip the targeted object before they sink in. At the same time, the toes splay and the paws become noticeably bigger. This is not only advantageous for the hunt, but also when used as a weapon during a fight, or to grip on to a tree when climbing.

Claws extend when the end bone of the toe, which bears the claw, pivots over the tip of the next bone. The claw is then retracted so that it doesn't touch the ground during movement, and the paws are soft once again.

Unsheathing the claws

The claw mechanism which makes a cat's claws extend and retract is very complex, and is adapted according to the situation: for example, when climbing, hunting, or fighting. It also depends on a cat's mood. If he is very upset, his muscles will tense and the claws are unsheathed automatically, and only retracted once the tension eases.

Also during general play or cuddling, the claws can be suddenly and unexpectedly extended. This is something which happens subconsciously, and is not intended to cause any harm, so to reprimand your cat for doing this would be completely

Claws like crampons: When coming down a tree, the claw mechanism doesn't work, so the cat either has to descend backwards ... or jump!

uncalled for, as he wouldn't have a clue what he'd done wrong.

Kitten claws need time to mature

The subtle interaction between the toes and muscles in a cat's paw doesn't function immediately from birth. It is not until the kitten is five weeks old that he can retract and extend his claws properly. When he kneads on his mother's stomach whilst suckling, he doesn't hurt her because his claws aren't yet properly developed. This also ensures that no harm is caused to the mother's birth canal when her kittens are born. Through growth and a gradual drying-out process, the claws harden, until after several weeks they have reached optimum consistency. The crescent-shaped claws also get sharper through use.

Kitten claws are soft and malleable, and at this stage they are not able to retract them.

Listening with the feet:

Sensory paws

The velvety-pawed pussy cat. When cats retract their claws between the soft paw pads, they can creep about without making a sound – to the misfortune of their prey ...

Can cat's paws 'hear'?

Cats' feet are capable of more than just scratching and sweating: they can also 'hear.' It's not that cats actually have aural receptors in their paws, of course, as they do in their inner ear; they are actually additional nerve endings which are found under the surface of the skin of the paw pads, at the base of the claws. These nerve endings are called the Pacinian corpuscle.

Because there are so many of these nerve endings – and they are so incredibly sensitive to vibration – the cat can sense the tiniest tremor, for example, made by a mouse underground.

The sensory information which comes from the paws is not only essential for hunting, but also useful for running, jumping, sure-footed climbing, balancing, and gripping, so these nerve endings are, in fact, essential for survival.

Striking out

You are bound to have seen your cat with prey (whether real or in the form of a toy), and noticed that he uses his paws to make a peculiar probing movement. The explanation for this is as follows.

Because the Pacinian corpuscles in his paws are so sensitive, the vibration effect rapidly wears off. Therefore, a cat cannot feel the vibration after the first sensory impression of the prey.

In order to sense further vibration, cats must 'reposition' their receptors, and this is only possible if they tap the ground with their paws to sense any further movement from the object.

Cat paws are lavishly constructed, all-purpose tools, which have many different, skilful applications.

Carpal paw pads and the 'vibrissae'

The little cone-shaped callous above the base of the forefoot is called the carpal paw pad. This provides extra touch sensitivity which helps with climbing and gripping prey. The fine sensory perception in the paw area also has three to six visible whiskers, which are found on the front legs, almost directly above the paw pads. These whiskers function as vibration receptors, and are called the vibrissae.

Paws for thought:

The vibrissae, whiskers, and other fur

It is not only the surface of a cat's skin that is sensitive: millions of 'touch receptors' – nerve fibres – can be found at the root of every tiny little hair.

A cat's strikingly high sensitivity from top to toe doesn't mean he is always sensitive to pain. Cats are very clever at hiding and protecting their sensitive areas

The whiskers

The impressive whiskers on a cat's face are specifically designed to sense touch. These whiskers are found around the nose and the mouth, and the touch receptor – or vibrissae whiskers – are mainly seen around the eyes, on the cheeks, the forehead, and the chin, as well as on the carpal paw pad. Some of the whiskers are spread out over the whole body – these are called the guard hairs, which a kitten has from birth.

If a whisker makes contact with something, the cat instinctively shuts his eyes to protect them from harm.

Hairs with improved stimulus receptors

Whiskers are basically modified body hair that are somewhat longer, stiffer, and have much deeper roots in the skin. Also, the network of nerve fibres are thicker at the roots, which means they have increased sensitivity.

At the base of each hair is something called a blood sinus; a little sack filled with blood which substantially improves the function of the stimulus receptors. So when the whiskers are moved, blood flow is directed to the base of the hair, which makes the nerve endings more sensitive. A sinus hair can move a minimum of 5 nanometres (2000th of the thickness of a human hair), which activates the already efferent nerve endings. An ultra gentle breeze would be enough to activate this reaction.

Better than night-sight

The tiniest current of air can be caused by prey, but can also indicate obstacles. These clever little hairs on the cat provide him with protection whilst he is on the move. They are especially useful when roaming through unknown territory, where it may be necessary to squeeze through narrow openings, or chase prey.

Thanks to these hairs (the whiskers

The whiskers are fanned out and pointed forward. They surround the grasshopper like a clasping hand. Since cats cannot see very well at close range, they use their vibrissae hairs to pinpoint the best position for a deadly bite.

on the face are the best indicators), it is possible for the little hunter to go out in the dark and move around safely without bumping into things, because the whiskers can sense anything that is in the way.

Bend and stretch

On average, cats have 24 whiskers over their upper lip (12 on each side) which lie in four horizontal rows. Like all other feline body hair, the whiskers can stand up on end or lay flat against the face.

When a cat comes across an object of whatever description, the whiskers stand on end and completely encircle it.

Cats can move one half of their face independently to the other, in order to test out surroundings. The rows of whiskers are able to tilt in different directions, which improves a cat's orientation abilities. Of all the whiskers, the vibrissae hairs are the most effective when it comes to steering the cat in the right direction.

At a glance:

A day in the life of a kitten

Fleur wanders off alone

Fleur, still young and inexperienced, is getting a good telling off from her mother because she has wandered too far. After a few months, a mother is happy to let her kittens roam around, and they will certainly take advantage of this eagerly-sought freedom.

Hunting in the meadow is great fun, but can be dangerous for a cat. In this situation, all the senses are on high alert, and the muscles are working in overdrive.

Oops! What was that ...?

Very young cats stay on their guard at all times when outdoors. When wandering around in open spaces, they will regularly glance above them: a hawk could be watching, just waiting to swoop.

Here, it is only Tinker, the tom cat, perched high in a tree who is watching.

Lilly stays close to her mother whilst Fleur, tail in the air, rushes to greet Lucy, the adult cat. But this effusiveness is an irritation, despite the fact that Lucy greets the little one appropriately and responds with a head-rub. A long-lasting friendship is certainly not sealed here, since Lucy is not a particularly sociable animal, although loves to cuddle up with Tinker, the neutered tom.

3

do you understand my body language?

All-round flexibility: What the whiskers tell us

Fear...!

The mood of a cat can be read by the position of his whiskers. Anxious cats flatten their whiskers against their face: the hairs squash together to form a band which points toward the back of the head. The lips are pressed into a thin line. These tactics are employed to make the face smaller, which shows submission to the enemy and could help avoid a potential attack.

Puffed-out cheeks

There is a completely different reaction when a cat is very cross. Below, the moggy puffs out his cheeks, making his face appear much bigger as he prepares to attack. The whiskers are fanned out and pointed forward over the mouth.

Rows of whisker hair are not only used as a warning sign but also improve a cat's capability for detecting prey or

Puffed-out cheeks and widely fanned, stiff whiskers testify to the 'bad atmosphere' here (as do the flattened ears and lashing tail).

danger. The whiskers are well-placed to guarantee a virtually 360 degree 'view.'

Completely relaxed

A relaxed cat holds his whiskers at the sides of his mouth rather than fanned out, and the upper lip and whiskers form a line. Sometimes the tip of the tongue pokes out of the mouth (or occasionally the entire tongue), which gives the cat a naive, childlike expression.

If something piques his curiousity, the whiskers are lifted in the direction of the object of interest. The whiskers will be far more spread out than when the cat is relaxed.

Sometimes the tongue remains poking out if the moggy is amazed at something he has just seen!

Concentration barometer

Simply put, how far the whiskers are fanned out can be regarded as a measure of how hard the cat is concentrating, or how attentivo hc is.

But this measurement is far more obvious if you have a camera handy since the human eye is not well trained or sensitive enough to follow the rapid and subtle changes.

Take the time to build a collection of photographs of your feline friend in all his different moods. With this, you'll be able to spot the tiny differences in the whiskers on both sides of his face.

Relaxed yet attentive, Lui strolls through his territory. His strongly expressive vlbrlssae hairs are only slightly fanned out, demonstrating his relaxed state.

Her interest awakened, this youngster tilts her ears forward, and fixes her gaze firmly on something. Note her whlskers are slightly fanned out and pointing forward also.

Impressive moves: Speaking with the ears

The most important mood barometer for cats are the ears. The different positions of the ear muscles indicate what frame of mind they are in

Little radar screens

Cats' ears are striking, no question about it., and the muscles are incredibly versatile. Cats can control their ears independently of each other, and they constantly change position.

The ears can be tilted forward and backward; rotated almost 180 degrees, and used to antagonise enemies. They can also serve as an unmistakable warning signal.

An astonishing 32 muscles per ear (we have just 6) alllow them to rotate, which not only helps with hunting and fleeing, but also plays a part in communication. By careful observation of the position of our cat's ears, we have a chance to learn something about our cat's mood and emotional state.

Deep in concentration; the ears point forward.

The language of the ears

If a cat is happy and relaxed, or if he has just had a nap, the ears tilt upward but are not tensed. The cat will carry them tilted slightly backward also, with the ear opening facing forward or very slightly out to the side.

If something has caught his attention, he will turn his head in that direction and point the tips of his ears toward the object of interest. In the picture on the left, the ear muscles are tensed and raised, the forehead muscles gently pulled inward (in the same way that we wrinkle our foreheads when we concentrate), and the ear opening faces forward. This also happens if the cat is eavesdropping! If the ears begin to twitch, this means he is stirring and ready to play.

Cantankerous or simply concentrating ...?

If the ear muscles turn outward so that a little of the back of the ear is visible, the cat is keyed-up or tense. It could be that he feels he has been disturbed or interrupted, so for a while is grouchy or annoyed.

Lots of cats use this ear position

The first indication of discontent is the ears pointing backward.

A defensive stance means the cat is irritated, and ready to fight.

when they are busy scratching their claws or checking scent-marks. It also signifies curiousity and concentration, for example, when scenting another animal's pheromones. Also, when cats clean themselves, the ears tilt slightly backward.

Still undecided

If the cat is unsure in any situation, trying to make a decision or feels ill at ease, he will prick up one ear and leave the other flat against his head.

Ears flat against the head

An anxious cat will flatten both ears against his head, and they will usually point backward as well. The greater the anxiety, the flatter the ears will lie against the head.

In the worst case, the ears are pressed right back against the skull to protect them from potential harm. This also displays submission so as not to provoke an opponent.

Warning! Raised hackles!

Aggression is what's usually being shown when the ears are turned inward as far as possible, so that the opening of the ear is at the side.

A cat that tilts his ears like this is often ready for a fight, although, conversely, it could also mean he feels on the defensive. In either case, a cat should not be hassled or provoked at this time!

Pure aggression: as well as the position of the ears, the open mouth (complete with wail) and whiskers pointing forward mean this cat is in no mood to play!

Ear gymnastics!

At just three weeks old, kittens are able to pinpoint the direction of different sounds with their ears. They display 'ear-acrobatics' in their first week as they learn to rotate their ears to different positions. It is a useful exercise for later life because it will help them to hunt successfully. By being able to locate their prey, half the battle is already won.

Look me in the eye little one!

Language of the eyes

The opponents stare each other down. One way of avoiding a full-on fight is to assert dominance by staring at each other in the hope that someone will back down.

Cats have a 360 degree view and can see movements out of the corners of their eyes

A declaration of war

Eye contact is an important part of a cat's body language. If his opponent holds his gaze for a long time and does not waiver, much like humans, he will begin to feel uncomfortable.

He will blink, look away, shut his eyes or quickly run his tongue over his lips (all of the so-called displacement activities which are also gestures of appeasement) in order not to provoke conflict. Or he will walk away to avoid the unpleasant situation. In the cat world, staring, or perhaps glaring, is a declaration of war.

Deeply affectionate gaze

A cat who is familiar with the eye contact of his human will sometimes hold this eye contact with a long gaze, occasionally accompanied by a 'silent' miaow. They understand our friendly intentions if we are staring at them, and we also understand their eye contact, which we know is not a glare.

If our cat is pleading for something, he holds his whiskers tense at the sides of his mouth, and while he gazes at us, his ears point upward and his pupils widen optimistically.

Far from funny – the menacing look

A lot more emotion is portrayed by a menacing look. As well as the penetrating glare, the ears are pressed back and turned inward (so that the back of the ears are visible from the front), and the whiskers stand out stiffly in front of the face.

This body language can sometimes be accompanied by furious hissing and narrowed pupils.

Pupil width

The size of the pupil is usually a sign of how light or dark the surroundings are. It can also be an indication of what mood your cat is in. If he is anxious or frightened, surprised or defensive, the pupils become very large.

With severe tension as well as pain, the pupils will narrow. If the cat is furious, his pupils will narrow to small slits. If he is pleasantly curious about something, as a general rule, the pupils will widen slightly, depending on the available light.

The light factor

The strongest indicator of a cat's emotional state Is not the actual size of his pupils but the sudden change in size from wide to narrow, or vice versa.

Even so, interpretation is not easy since light and arousal influences affect pupils by either exaggerating or underplaying the reaction.

A happy, relaxed moggy will widen his pupils only so far as he needs to to let in the light (a slight oval shape). They are never really noticeably wider than this in daylight because cats only need a sixth of the light that we do in order to see.

Cats' eyes phenomena

At dawn or dusk, the pupil widens. The darker it is, the bigger the pupil, in which case the pupil is shaped like a circle and takes up nearly the whole surface of the eye. The eye then looks as though it is completely black. This reflecting reaction draws in as much light as possible to enable a cat to see in poor light. This happens to the detriment of visual acuity because the effects of scattered light cause the picture to be obscured. In bright light, the pupils narrow: the stronger the light, the smaller they become. In glaring sunlight, they narrow to a tiny vertical slit because the cat has a photosensitive retina which protects the eye.

This cat is completely relaxed and at ease: the eyelids are half-open and the nictitating membrane (the third eyelid) is visible.

For the kids:

What your cat's face is saying

You can read your cat's face like a book. Don't just look at his ears, eyes and whiskers on their own. though, but take all of these into account at the same time, and this will tell you how your cat is feeling.

Chilled out

If a cat is completely relaxed, his ears point upward, with the opening facing forward. His whiskers sit at the side of his cheeks, or point slightly in the direction of the chin. He looks at you in a friendly way, but the size of his pupils will depend on how light it is.

Warning! Moody moggy!

If a cat is annoyed, he pulls his ears flat against his head, his whiskers stick straight out in front of his face, and his pupils are narrowed. This tells you that your cat would like to

be left in peace so don't disturb him, otherwise you may get scratched!

Fear

Eyes wide open, enlarged pupils and ears flat aginst the head tell you that your cat is scared. His whiskers may also be pressed flat against his cheeks. Do not tease or annoy him under any circumstances, as his fear could soon turn to anger!

Come and play with me!

If a cat is in a playful mood, the tips of his ears point upward. The forehead appears much higher than usual, and the pupils are dilated with excitement. The whiskers point forward, though are not stiff as with an angry cat, but softly arched instead.

Your cat may even let out a little

miaow or prod you with his paw to encourage you to play with him.

Child's play for cats

Take a toy (for example, a mouse tied to a piece of string), position it in front of your cat and then make it 'dance.' Your cat will follow it with interest and reach closer and closer to try and hook it with his claws.

Don't play this game with bare hands, though, because you could get scratched in the heat of the moment (not on purpose but by mistake), and the game might come to a sudden end. You definitely don't want that to happen!

Dog-tired

If your cat is tired and blissfully content, he will curl up in a warm, snuggly place with his whiskers relaxed, the third eyelid (which we do not have) extended over his eye, and his eyes more or less completely shut. Don't disturb him – let him doze! Whilst he is 'recharging his batteries,' you will notice that he snuffles his little nose and his ears tilt upward.

Perseverance is required:

Moggy on the hunt

Lying in wait

The excited twitch of the tail signifies that this cat is about to go on a hunt.

The body assumes a crouching position, and the eyes are firmly fixed on a hollow in the grass: every movement is smooth, subtle and deliberate.

The hunter crosses the open meadow. Momentarily distracted, she freezes before lifting her velvet paw and stepping over the uneven ground. Slowly, very slowly, she places her paws on the ground so as not to disturb the grass. For a moment, she holds her position, completely absorbed by what the vibrations in the ground are telling her.

Scanning for prey

With toes now splayed, the cat begins to carefully rub and touch the grass with her paws, and gently moves forward.

Her pupils are noticeably larger, ears are tilted forward, and whiskers point straight out in front.

You can almost feel the tension in the air ...

Watch out!

Suddenly, she springs forward and lands with her back tensed in an arch and claws extended. No luck this time, though ...

Disappointed, she sniffs the blades of grass and then moves away. Even though she didn't catch anything this time, it's all good training for her muscles and senses.

A day's hunting

On the hunt for prey, cats roam the area with senses fully alert. From time to time, they will stop and freeze in order to scan their surroundings, using not only their eyes but also their nose and whiskers.

If they sense something interesting, they fix their gaze firmly on it and creep toward it. This is known as stalking.

Patience at the mouse hole

In a well-known and familiar environment, the hunt is a little easier.

Rather than stalk their prey, cats prefer to lie in ambush. For example, they may specifically target a mouse hole.

This 'sit-in' type of hunt can last a while – half an hour is not unusual. Cats are able to exercise great patience. They carefully sit in position in front of a hole, and wait, tensed, for the first sign of movement.

Their ears will detect any sounds (cats respond especially to high frequencies) as their eyes flicker and scan the area, whilst the nose sniffs and snuffles incessantly. The vibrissae hairs sense every waft of air.

Prey spotted, he pounces with claws fully extended! The vibrissae hairs help to locate the exact point for the deadly bite in the mouse's neck. Close-up, a cat's vision is not as good and he is not able to perceive depth and distance very well.

Teeth chattering

If a cat sees his prey but is unable to get to it (a bird high in a tree, for example), he will sometimes make a strange bleating, chattering noise, where the mouth is slightly open, the lips are pulled back and the jaw quickly opens and closes.

This chattering is a displacement activity which is probably used to relieve tension and frustration.

The 'sit-in' hunt – his patience knows no bounds ...

A typical example of 'chattering' – because the prey is out of reach!

Crash course in hunting

An adult cat will usually hunt alone. But a mother cat will go hunting with one or more of her kittens in tow so that they can learn hunting skills by watching her expertise.

From fun to rough and tumble: Cat play

Playing with prey

With his catch between his teeth, a cat may trot over to a secure and comfortable spot to eat his meal in peace. Sometimes, though, the catch is not eaten.

Often, cats will grip the prey with their claws to hold it still, but do not kill it straight away. Instead, they allow the creature to move a little so that they can then swipe it with a paw or toss it into the air, a behaviour that is both cruel and playful at the same time.

This behaviour, presumably, helps a young cat to become more co-ordinated, whilst at the same time perfecting his hunting abilities.

It has been scientifically proven that hunger and the desire to hunt originate in different areas of a cat's brain, and these areas are not used in direct relation to one another.

This means, of course, that cats may hunt when they are not hungry.

In this instance, the cat may simply leave the dead prey without attempting to eat it, or, if it's still alive, allow it to escape only to catch it again to keep the game going.

This little kitten is already honing his skills in a game with one of his siblings.

Quick and snappy

Cats love quick movement with abrupt changes in direction, either when playing with each other or with people.

The mother's twitching tail is an Invitation to the playful kitten to begin jumping around and chewing on it. Cats playfully tap each others' faces with their paws, but this can quickly get out of control and end up in a real clobbering: a good boxing of the ears, lots of angry hissing, and then they start to chase one another. Before you know it, the fun is over and a proper scrap is in full flow.

A sudden change of mood

During play, certain skills are honed which will be needed for hunting, fighting and breeding. Cats are very independent and have instincts which are not geared toward living in a big pack with a graded social hierarchy.

So it is easy to see why games with a friendly motivation can suddenly turn serious. In spite of this, there are subordinates to the dominant cats who live in a group long after they have fully matured and get along very well, so it really does depend on a cat's nature.

Some cats prefer to groom each other rather than playfight because they are fully aware of where this can lead.

For training purposes

How cute! Pouncing on a piece of string, chasing after a ball, or jumping high in the air for a treat is not only a part of play, but a natural element of hunting and fighting behaviour. Our moggies know how to improve their reflexes, balance, speed and movement through training, *and* have fun at the same time!

In a pack which gets along well, adult members love to play together. Most games are quick and frequent.

Balance is the trump card: No fear of heights

Cats develop fantastic climbing skills, and their sense of balance is remarkable.

Safe landing

Slow-motion recordings show that when a cat falls from a great height, he raises his head, turns his face toward the ground and lifts his hindquarters so that he brings his whole body into horizontal alignment. The tail serves as a type of rotor which also helps with balance. Next the cat brings his legs [illegible]

Child's play

Hot-footing it up the nearest tree, a sprint across flimsy branches, or walking the tightrope over narrow fences – all of these are easy for cats.

With lithe body, whiskers fanned out and agile paws, they climb quickly and softly, and show off their skills with every movement. Cats have a remarkable sense of balance which allows them to master the tightrope walk in a way we can only dream of.

A balancing act

A cat's sense of balance is due to an extremely efficient organ in the inner ear. This sensitive organ also makes it possible for a cat to stabilise himself after jumping or falling from a great height, turning his body in the air to land on all four paws so that he does not hurt himself, like a dog would if he fell from the same height, for example. Of course, gravity usually dictates that the heaviest part of the body lands first.

Learning how to fall

This specific reflex is not evident at birth but develops in the sixth week of life. Up until this point – and actually for a fairly long time afterward – a kitten does not have a very precise perception of depth because his spatial ability is not yet properly developed. Also, his little tail is clumsy and not yet trained as a balancing tool. The tiny claws are still quite soft, which also affects how well he can balance. As the kitten is so vunerable, this is quite a dangerous time for him, so he will tend not to stray too far from his mother, relying on her to keep him safe from harm.

Preparation begins during gestation

A sense of balance is developed long before it is actually perfected by the cat. At the time of birth, the highly sensitive cells of the balance organ are already working at full capacity. This means that the tiny kitten can pick up balancing skills very quickly. Without an impeccable sense of balance, the cat would be doomed.

Interestingly, this sense of balance stays nearly as precise right into old age.

Senior cats

A very old cat's sense of balance will decrease slightly, and the animal may sometimes trip over or appear unsure about balancing acts that he once found easy.

Also, his other senses are somewhat diminished. This is when an owner should try and adapt their surroundings to better suit their senior feline friend.

Going up is much easier than coming down. The little tom cat is a bit reluctant to jump. He will have to come down backward until his balance improves.

Mood barometer: Speak to the tail

If a cat flicks his tail, it can indicate either emotional conflict or a strong interest in his prey

The tail acts as a counterpoise and helps the cat to keep his balance. But that's only half the story.

The cat's tail is also an impressive communication tool, and what he expresses is not only understood by other cats but by people as well. The position and movement of the tail means it is possible to read what he is actually saying.

Content and good-humoured

A relaxed, content and alert cat who is strolling through his territory holds his tail loosely in the air, curved slightly to one side with the tip pointing upward. If something catches his eye, he straightens his tail.

If the tail is in the shape of a question mark, the moggy is in a great mood and eager to explore.

Here, the cat is scent-marking his territory.

Decisions, decisions

If a cat encounters something that needs to be weighed up and considered, his tail will first begin to twitch. and then flick back and forth, indicating that he is completely torn about which decision he should make.

It may be that he has scented prey that is slightly threatening, but he is curious, nevertheless. This tail twitching is definitely not a sign of contentment, but neither is it a sign of aggression.

The cat is experiencing emotional conflict. As soon as he has decided what to do, his tail will stop twitching.

Menacing aggression

If the tail begins to lash from side to side, this is a display of agitation. The cat is extremely irritated, and this could quite easily turn to aggression. An attack is a distinct possibliity.

A lamb's tail

The so-called lamb's tail is a sign of conflict: for example, mistrust or

uncertainty. In this instance, the tail sticks out at the root and then drops straight down. If it should also twitch, and the animal has raised hindquarters, this is not a good sign. The cat is very anxious, and is also displaying threatening or defensive behaviour.

A bushy tail

When feeling under threat, the hairs on the tail stand up on end. If ready to fight, all the hairs along the cat's backbone stand up and the tail – which now looks like a bottlebrush – points straight out. The cat uses every trick possible to appear bigger and more threatening to his enemy.

Fear and shock

If a cat is fearful or shocked, (this also occurs if the cat is defending himself) every hair on his body stands on end.

The tail held in a curve over his body is a display of aggression; if the tail is held downward, this means he is afraid. To show submission he will sometimes tuck his tail between his back legs, just as dogs do.

Hello, my friend!

If the cat meets another friendly cat or his owner, the tail is momentarily raised in the air to give his companion the chance to sniff his rear end to enhance the bond between them. Following this, they will rub their tails together and then over each other's back in order to exchange scents. Sometimes when cats greet people, the equivalent welcome is to wrap their tails around the person's legs.

Top: The lashing tail of the young tom cat shows that he is irritated, but his mother keeps her cool, although her ears are very flat on her head.

Above: The raised, entwined tails of both cats is a sign that all is well.

Body language:

Arching the back and other gestures

This little kitten has already learnt to display the back arch, which can signify fear, upset or cockiness.

A cat who is pleased will 'grow' taller. A cat who wants to hide makes himself as small as possible

Arching the back

Young kittens arch their backs when playing together, sometimes accompanied by ruffled-up fur. They may also 'hop' sideways around the room. It looks like they are being cocky but this behaviour can also signify a hint of fear.

Back arching also occurs when a cat prepares to defend himself in a fight that he is not confident about. A cat who feels threatened raises himself up on his forepaws and the hair on his back stands on end. He is demonstrating his position of attack, but at the same time gently pulls his forepaws back and curves his back into an upside down U-shape (a gesture of fear).

WIth a friendly back arch a cat will stretch his legs; all of the hairs lie flat against the body and there's no muscle tension in the hindquarters.

Getting taller

Cats who want to take an enemy by surprise make themselves as tall as possible. They stretch their limbs, raise their body as high as they can, and ruffle their fur in the hope of misleading their opponent and avoiding a potential fight. Since their hind legs are longer than their front legs, when they raise

A self-assured and confident kitten: his little tail is raised in the air as a greeting.

themselves up, their bottom is higher than the neck.

In general, stretched legs and a raised bottom and head signify security and dominance, as well as preparedness for attack.

Aggressive cats make the fur on the spine stand up on end and curve their bristly tail so that they appear more imposing. They're going all-out to frighten their opponent by their size and confidence.

Shrinking

Very anxious, shy or submissive cats shrink into themselves to look as small as possible so as not to provoke an attack from an enemy. Or they may simply be uncertain of something or someone.

As they slowly creep away from the danger – head down and avoiding eye contact – they make themselves as flat and long as possible so that their stomach almost touches the ground. By taking this action, it is hoped to avoid confrontation.

This cat looks on anxiously from her discreet hiding place.

At a glance:

Feline facial expressions

Completely relaxed

This is a relaxed cat who is surveying his surroundings. His eyes are fixed on something, and the eyelids are open and pupils dilated just enough to let in the right amount of light. The vibrissae hairs are soft and held in a relaxed position.

Where is the grasshopper?

If something has caught a cat's attention, muscle tension increases. Ear openings face forward, the eyes and pupils widen, and the whiskers stick straight out in the direction of the object of interest; in this case, a grasshopper. A cat will fix all his senses and concentration on his prey before pouncing.

I'm harmless

A typical gesture of appeasement: the ears are pricked, the eyelids slightly closed, the pupils become small, and the cat blinks. The whiskers remain grouped and are stiffly raised to form a band that is as inconspicuous as possible, and the tongue comes out to lick the lips.

This cat is anxious, uncertain

The puffed-out cheeks of this virile tom cat are the result of his sex hormones, and not an expression of threatening behaviour.

Tired tiger

Yawning to her heart's content: this moggy is feeling relaxed and playful. She is happily sprawled inside an empty window box. Even haughty felines sometimes play the fool!

or nervous, and doesn't want to provoke his opponent in any way. In this situation, cats may also make themselves appear smaller by holding their bodies stretched out and close to the ground.

Puffed-out cheeks and big eyes

Large dilated pupils can indicate fear or upset, but if the ears are pointed upward at the same time (as here), this means a cat is more attentive than fearful. In general, upset or lack of light at dawn or dusk can cause the pupils to dilate.

Cat fight

Staring each other out, ears pulled back, rigid whiskers: anger is in the air! These cats will either creep past one another or have a full-on scrap.

Further reading

Cat Body Language Phrasebook: 100 ways to read their signals
by Trevor Warner. Thunder Bay Press (2007). ISBN: 9781592237104

The Body Language and Emotion of Cats
by Myrna M Milani. William Morrow; reissued edition (1993). ISBN: 9780688128401

The Cat Whisperer
by Claire Bessant New Edition. Blake Publishing (2003). ISBN: 9781904034742

How to Say it to Your Cat: Understanding and Communicating with Your Feline Illustrated Edition
by Janine Adams. Prentice Hall Press (2003). ISBN: 9780735203297

Cat Psychology & Cat Behavior Tips: Ultimate Cat Advice Guide
by Julie Seymore. Kindle version. Amazon Media EU Sarl. ASIN: B005D1OSBS

Fun and Games for Cats
by Denise Seidl. Hubble and Hattie (2011). ISBN: 9781845843878

Miaow! Cats really are nicer than people!
by Sir Patrick Moore. Hubble and Hattie (2012). ISBN: 9781845844356

How To Talk To Your Cat
by Claire Bessant. John Blake Publishing Ltd (2008). ISBN: 9781844545155

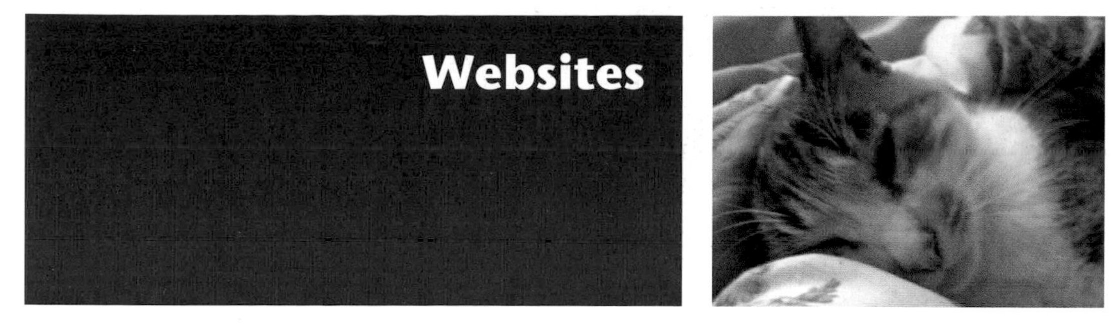

Websites

http://en.wikipedia.org/wiki/Cat_communication
Information on miaowing, purring, biting, scent, and body language – and what it all means!

http://www.knowyourcat.info/index.html
A great website dedicated to all things cat, including health and body language

http://www.best-cat-art.com/cat-body-language.html
Information and facts about cat body language

http://www.messybeast.com/cat_talk2.htm
Cat communication, including feline vocalisation and the language of smells

http://animal.discovery.com/cat-guide/cat-behavior/reading-your-cat.html
Cat behaviour and body language, how cats purr, plus lots of useful links for cat enthusiasts

http://cats.about.com/od/behaviortraining/ss/bodylanguage.htm
Useful website showing pictures of different types of feline body language

http://www.cat-behavior-explained.com/cat-body-language.html
Cat behaviour and body language explained

http://www.a-house-full-of-cats.com/catbodylanguage.html
Cat body language: talk to the tail!

http://www.cat-training.co.uk/cat-body-language.htm
Miaowing, body language, scratching, spraying, biting and much more

http://www.paws-on-floors.co.uk/bodylanguage.html
Feline body language, diary of a cat, cat rules, and cat dreams

http://www.thepurrcompany.com/cat-articles/index.php?id=60
Cat psychology and behaviour – read your cat's body language!

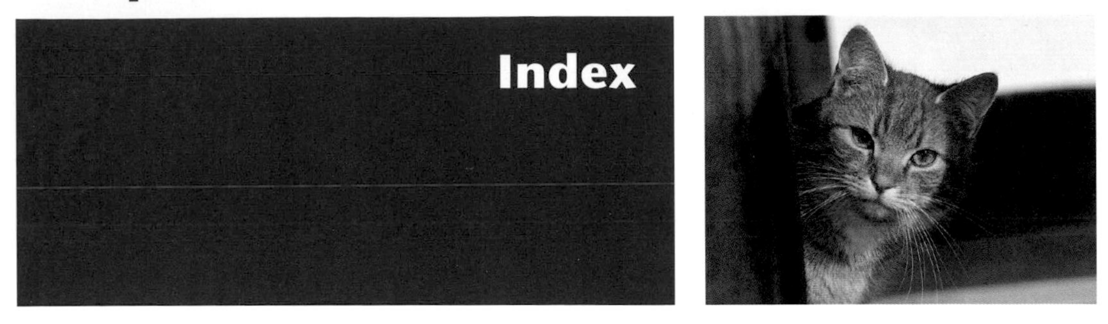

Index

"I chose the title for my book quite deliberately. Obviously, it is dangerous to generalise, but given a choice between the average person and the average cat, I would opt for the cat any time, and in this book I will try and explain why.

Here goes ..."

... dear Jeannie ...

... beloved Ptolemy ...

... how I loved cricket! ...

... Ptolemy with me in my study ...

For more info about Hubble and Hattie books, visit www.hubbleandhattie.com • email info@hubbleandhattie.com • Tel: 01305 260068 • *prices subject to change • p&p extra

22x17cm, 64 pages, 35 colour pictures, paperback, ISBN: 9781845844356, £7.99*